D. C Lloyd-Owen

The Elements of Ophthalmic Therapeutics

Being the Richard Middlemore Post-Graduate Lectures...

D. C Lloyd-Owen

The Elements of Ophthalmic Therapeutics
Being the Richard Middlemore Post-Graduate Lectures...

ISBN/EAN: 9783337022082

Printed in Europe, USA, Canada, Australia, Japan

Cover: Foto ©berggeist007 / pixelio.de

More available books at **www.hansebooks.com**

THE ELEMENTS OF

OPHTHALMIC

THERAPEUTICS,

BEING

The Richard Middlemore Post-Graduate Lectures

DELIVERED AT THE

BIRMINGHAM AND MIDLAND EYE HOSPITAL, 1889.

BY

D. C. LLOYD-OWEN, F.R.C.S.I.,

SENIOR SURGEON TO THE HOSPITAL, AND CONSULTING OPHTHALMIC
SURGEON TO THE CHILDREN'S HOSPITAL, BIRMINGHAM.

BIRMINGHAM

CORNISH BROTHERS, 37, NEW STREET.

1890.

CONTENTS.

INTRODUCTION.

———

I N delivering the first of the annual series of lectures established by Mr. RICHARD MIDDLEMORE, I am privileged in being able to act as the mouthpiece of the medical profession in this district, in expressing their cordial appreciation of his generosity, and of the service done to his professional brethren in founding and endowing these lectures.

Although Mr. MIDDLEMORE has reached a very advanced age, he is well known to take a warm interest still in the progress and advancement of medical science, and especially of ophthalmology, to which his life has been devoted.

From his earliest connection with this branch of medicine, which now dates back some sixty-odd years, Mr. MIDDLEMORE has always laboured, by his lectures and his writings, to place the results of special research before his brethren in general practice,—a high-minded line of conduct which has now culminated in the establishment and endowment of the " RICHARD MIDDLEMORE Post-

Graduate Lectureship," in connection with the early scene of his labours, the Birmingham and Midland Eye Hospital. This unity of method and purpose manifest throughout a long life, is a fitting illustration of the poet's description of a worthy and complete life :

"Une pensée de la jeunesse, exécutée dans l'âge mûr."

And I congratulate Mr. MIDDLEMORE, in the name of our profession, on witnessing the completion of his work.

Mr. MIDDLEMORE is a lineal descendant of the Hawkesley branch of the ancient and distinguished family of MIDDLE-MORE, formerly lords of the manor of Edgbaston, and possessed of other manors and great estates in the counties of Warwick and Worcester : a family of great importance from the time of HENRY II down to the time of the early GEORGES.

Mr. MIDDLEMORE was a student of St. Bartholomew's Hospital, where his chief friend and companion was Professor, now Sir RICHARD OWEN, K.C.B., a friendship which has continued warmly cemented to the present day. On leaving St. Bartholomew's, Mr. MIDDLEMORE came to Birmingham the bearer of a highly commendatory letter from Mr. ABERNETHY to Mr. HODGSON, then Surgeon to the General Hospital here, and by far the most distinguished surgeon throughout the midlands. For three years Mr. MIDDLEMORE acted as dressing pupil to Mr. HODGSON, and as his friendly assistant for more than ten years afterwards. Instigated probably by Mr. HODGSON'S example, Mr. MIDDLEMORE early evinced a leaning towards the study of ophthalmology, and in the year 1828 he was appointed assistant surgeon to the Birmingham and Midland Eye Hospital; in 1835 he was made full surgeon, and in 1849 he retired to the consulting staff. In 1831 Mr. MIDDLEMORE gained the Jacksonian prize; in 1834 he delivered his first course of lectures on diseases of the eye; and in 1835 published his well-known work, *A Treatise on Diseases of the Eye.*

In addition to this great work Mr. MIDDLEMORE endeavoured to start a journal of ophthalmology, but the enthusiasm necessary was lacking among the ophthalmic practitioners. He is also the author of innumerable papers published in the various English, and quoted by the various foreign medical journals. By means of which it is known that Mr. MIDDLEMORE contributed greatly to the foundation of the English school of ophthalmology.

Mr. MIDDLEMORE was one of the first four members of the Royal College of Surgeons who were created honorary fellows, and is now the only survivor of the four; the others being well-known and honoured Birmingham men :

JOSEPH HODGSON, W. SANDS COX, and LANGSTON PARKER. Besides the establishment of the lectures which bring us here to-day, Mr. MIDDLEMORE has endowed a triennial prize in connection with the British Medical Association, and has given largely to establish a fund in connection with the Blind Asylum in this city, to help those who, having been educated in the institution, may be hindered in beginning a useful life by want of pecuniary means.

Long may Mr. MIDDLEMORE be spared to enjoy the fruit of his good works. Conscientious in his relations with his professional brethren : modest, never seeking notoriety, simple, kind, generous, always maintaining a high standard of professional life, Mr. MIDDLEMORE has furnished us an example which we may well follow. Of him it may be said, in the words of CICERO :

" Memoria bene redditae vitae sempiterna."

Chapter I.

In taking as the subject of these post-graduate lectures some of the essential points of ophthalmic therapeutics, I am endeavouring to follow, to the best of my ability, the intention of the benevolent and enlightened founder of this course : namely, to digest and arrange for the service of those members of our profession who are engaged in the absorbing duties of general practice, the results of the researches and experience of those who devote themselves more particularly to ophthalmology.

It is most probable that at least for some time to come the cultivation and mastery of the more abstruse details must be the pursuit and remain the possession of a comparatively small number of practitioners. But, leaving these more special aspects aside, there remains a large number of the less serious diseases of the eye which ordinarily come under the care of the family medical attendant.

The treatment of these diseases has been very carefully studied by ophthalmologists, and the rules which guide them are among the rudiments of their knowledge. But these rules are not yet as they should be the common possession of the whole profession. This I say without for a moment wishing such shortcoming to be considered due to causes other than the demands on the student's time, which leave him but little for the study of ophthalmology.

My lectures will be directed particularly to the treatment of disease; but as it is of little value to discuss remedies without at the same time endeavouring to make plain the indications for their exhibition, I shall try, when crucial points between diseases liable to be mistaken for one another demand, to add to the directions for treatment some hints for guidance in diagnosis.

I would say further, that it is not my intention to quote authorities in these lectures; and I would ask you to accept the suggestions laid before you as expressions of the general concensus of opinion held by ophthalmologists.

For convenience of reference I have broken the text of my lectures up into chapters and sections when producing it in book form.

___

SECTION I.

MYDRIATICS AND MYOTICS.

MYDRIATICS.—In the modern treatmeat of diseases of the eye, mydriatics and myotics have come to hold a very prominent place. For besides their action in dilating and contracting the pupil, and so affecting the merely mechanical adaptation of its changes to the requirements of treatment, they have an important influence in modifying the conditions of the nerves and blood-vessels of the eye, and so are eminently serviceable in the general treatment of its diseases. This latter feature in the treatment of eye diseases is of distinctly recent development; some of the drugs employed in its operation—notably cocaine—being of very late adoption.

I will take first of all the mydriatics—the dilators of the pupil—seriatim, and then compare their action and their therapeutic value.

The mydriatics generally employed are the vegetable alkaloids derived from plants of the natural family Solanaceæ; and they are all very similar, almost identical, in their chemical composition, and their therapeutic action. They are *atropine*, derived from Atropa belladonna; *duboisine*, from Duboisia myoporoides; *hyoscyamine*, from Hyoscyamus niger; and *daturine*, from Datura stramonium. These mydriatics from the solanaceæ, consist of only three pure alkaloids—atropine, hyoscyamine, and hyoscine. Daturine is nearly pure atropine, and duboisine nearly pure hyoscyamine, and all three alkaloids are isomeric. Besides these four derivatives from the solanaceæ, there are two other alkaloids, gelsemine, derived from Gelseminum semper virens, and cocaine, derived from Erythroxylon coca. The last, though more generally used for its anæsthetic effect, is also a valuable mydriatic, though the dilation produced is of brief duration. Besides their action in dilating the pupil, these drugs also affect the accommodation of the eye. That is to say, the intra-ocular muscles, the sphincter of the iris and the ciliary muscle, are both influenced by the action of mydriatics, the effect of the drugs being produced on the intra-ocular ganglionic system which governs them. It is a question not yet finally decided whether, besides paralysing the sphincter of the pupil, mydriatics irritate the oculo-pupillary branches of the sympathetic, and by so doing, add to the paralysis of the sphincter an irritation of the radiating fibres of the iris.

And this need not occupy us further now, than to notice that dilatation of the pupil produced by mydriatics can be carried to a more complete degree than is possible by interruption of the nerve supply from the third pair alone, without the addition of irritation of the cervical sympathetic.

To produce moderate dilatation of the pupil, an exceedingly feeble solution of atropine suffices; one part in eighty thousand will at the end of an hour produce a dilatation traces of which will remain for twenty-four hours. But in order to affect the power of accommodation a solution of at least one part in twelve hundred is needed. These solutions apply only to eyes in a normal state, they would not affect eyes the seat of disease, nor the eyes of some old persons in whom 'it is difficult to produce dilatation. An iris in a state of inflammation will not dilate in answer to a weak solution, nor will a ciliary muscle in a state of spasm yield readily. Strong solutions, and often many days of persistent use are necessary, before, in such conditions, the desired effect is produced.

The sulphate of atropine is the salt most generally employed, and it has the advantage of being readily soluble in water without the aid of spirit of wine or acid. The pure alkaloid, on the other hand, needs a small quantity of spirit of wine or acid to render it soluble in sufficient quantity to be of use, and it should not therefore be used as a local application to the eye, on account of the smarting and irritation produced by these agents, when brought in contact with the cornea and conjunctiva.

A solution of the pure alkaloid, however, in castor oil or olive oil, to the value of which I was, I believe, the first to call general attention, forms a very valuable local application in abrasions and ulcerations of the corneal surface, and in the irritation due to granular lids. Here the lubricating effects of the oil are gained, in addition to the sedative effects of the drug. The strength of the aqueous solution of sulphate of atropine most generally useful is that in the proportion of one in two hundred of water, but where the full effect is needed, as in producing extreme dilatation, in the attempt to separate adhesions of the iris to the anterior capsule of the lens, or to cause complete paralysis of the muscle of accommodation, a solution of one in a hundred, or even one in sixty, is required.

In certain persons, atropine acts as an irritant to the

conjunctiva, and in some, it will even cause redness and puffiness of the eyelids, and more or less eczema of the skin in the neighbourhood of the eye. When this occurs, it is of little use to continue its application. Sometimes atropine may be tolerated by substituting an ointment for a watery solution, or by adding to the watery solution a small proportion of carbolic or boric acid, or by using gelatine discs charged with the alkaloid. But I am bound to say, such a result is the rare exception.

The susceptibility to this form of irritation, I believe to be an idiosyncrasy, and it is better not to lose time in trying these variations of vehicle, but to at once change the drug for some other similar in action. The sulphate of duboisine or hyoscyamine are the best substitutes.

Duboisine possesses a much more powerful mydriatic action than atropine: a solution of one part in a million will produce marked dilatation of the pupil; a solution of one part in two hundred is as strong as is needed for ophthalmic purposes. It is especially worthy of notice, however, that whilst duboisine does not often act as a local irritant, it possesses marked toxic properties, and its use should be carefully watched.

Hyoscyamine has great value as a mydriatic, a solution of one in three hundred will fully dilate the pupil, and completely paralyse the accommodation, and the effect will not pass off for some days. The solution, however, does not keep well, and it is therefore less fitted for use, except in special cases, than the other mydriatics.

Daturine, which is almost identical with atropine, has no special value.

Within the last few years, a drug called homatropine, a synthetic salt, has been introduced into ophthalmic practice. It is a product obtained from the amygdalate of tropine, a salt derived from atropine by splitting it up into tropine and tropic acid, and substituting amygdalic acid for the tropic acid. It is obtained in the form of an oleaginous liquid ; and this, with hydrobromic acid, forms a soluble crystalline salt. This result, the hydro-bromate of homatropine, is used in ophthalmic therapeutics, and produces effects similar to those of atropine, but it acts more quickly, and its effects are of shorter duration. It is, therefore, very useful for diagnostic purposes, because whilst it produces full dilatation of the pupil, and complete paralysis of the ciliary muscle, its effects pass off in from twenty-four to thirty-six hours.

No drug has attracted more attention during the last three or four years than *cocaine*. It was only brought before the medical world by Köhler, of Vienna, in October, 1884 ; and before the end of the same year, I have counted between eighty and ninety communications to British and foreign journals, concerning its effect on the eye alone. Whatever may be its ultimate position in other divisions of practice, it is a settled fact, that it is most valuable as a local anæsthetic and as a mydriatic. In the latter role it is useful because it produces a brief but extremely wide dilatation of the pupil, sufficient to allow of the freest ophthalmoscopic examination, without more than the briefest disturbance of the accommodation, and with probably less danger to eyes threatened with glaucoma than is offered by other mydriatics whose effects are more lasting.

With homatropine cocaine forms a most useful combination, which will thoroughly paralyse the accommodation, whilst the effect does not last more than thirty-six hours.

The directly therapeutic effects of mydriatics are—

1.—To dilate the pupil, either to improve the vision, or to keep the iris free from adhesions to the capsule of the lens, during attacks of inflammation of the iris, and also from adhesions to the cornea, in ulcers or wounds.

2.—To relieve spasm of the accommodation.

3.—To relieve pain and reflex spasm by their directly sedative action on the terminal branches of the ocular nerves, when these are implicated in disease.

4.—In addition to these actions, common to all the mydriatics in greater or less degree, must be added the special local anæsthetic effects of cocaine.

To take the uses more in detail, I will first of all consider the simple dilating effect of mydriatics. This is employed when central vision is obstructed by some opacity in the media, such for example as a central, senile cataract, or a lamellar cataract of moderate extent. It is useful by permitting the patient to take the benefit of vision through a portion of the transparent media, outside the area of obstruction. For this purpose a very weak solution is all that is required, and even this is necessary only at long intervals, and should be used with caution.

Mydriatics are also used to draw the iris backwards towards its ciliary attachments, and prevent its incarceration in ulcers or wounds near the centre of the cornea.

To obtain and maintain dilatation of the pupil in iritis, or when post-iritic adhesions exist, strong solutions are necessary. If atropine be used, it must be in the proportion of five to eight grains to the ounce of water. In using mydriatics in acute inflammation of the iris, frequent instillations are necessary. It may readily be understood that in the treatment of inflammation of a structure like the iris, which is constantly and rapidly varying the position of its relative parts, in response to the stimuli of light and accommodation, it is before all things necessary that rest should be secured. Dilatation of the pupil to its fullest extent effectually insures the repose needed, and at the same time, by withdrawing the iris from its normal contact with the lens capsule, it removes the serious danger of adhesions between the two structures.

A very good way of insuring absorption, and of obtaining somewhat of a cumulative effect of atropine, is to instil the solution every five minutes for a quarter of an hour, and to repeat this proceeding every two, three, or four hours, according to the acuteness of the attack. This plan has two great advantages : first, it is more likely to insure absorption, because an eye in an inflamed condition, with its vessels gorged and its tension somewhat increased, will not absorb a fluid from without so readily as it would do under normal conditions ; and, second : even in ordinary cases there is excessive lachrymation, which will wash out the solution instilled, before time for absorption has been allowed. The repetition of the instillations every five minutes for a quarter of an hour, meets these adverse conditions better than any other method. They have another advantage, too—these quickly repeated instillations with good intervals—they are less disturbing to the patient, and less irritating to the eye, than instillations every hour or half-hour, as recommended by some authorities.

In attempting to break down iritic and post-iritic adhesions between the iris and the lens capsule, it is necessary to obtain full paralysis of the sphincter of the iris.

The existence of a distinct dilator muscle is still disputed, but some facts have been observed which, I think, show conclusively that paralysis of the sphincter is not all that may be reasonably hoped for in attempting to break down synechiæ.

There can be no doubt that atropine, besides paralysing the branches of the third nerve, stimulates the branches of the sympathetic and its associate the fifth, which preside over dilatation of the pupil ; and so excitement of these dilating nerves may be fairly reckoned upon as a factor of some value, in detaching the adhesions.

I would now draw attention to a most important caution in the use of mydriatics, and that is, that they must on no account be used in cases of glaucoma, or of threatened glaucoma ; that is, in eyes the tension of which is above normal. I lay particular stress upon this, because it happens now and then that we see cases of glaucoma which have been mistaken for iritis, and in which atropine has been used with disastrous effect. In order to explain this I will glance briefly at the pathology of glaucoma, and it will be seen in what way mydriatics are injurious.

Glaucoma is, briefly, an increased tension of the eyeball, due to the retention within it of the fluids secreted, through interference with the normal channel of outflow ; just as normal tension is the expression of a fairly strict equilibrium between the constant secretion of fluid into the eye, and its equally constant and proportionate withdrawal. By far the most important route taken by the escaping fluid is that through the anterior filtration area, as it is called ; that is, through the ring of tissues immediately posterior to the corneal margin, and anterior to the ciliary margin of the iris, marked by the ligamentum pectinatum. This ring marks the site of Schlemm's canal, with its contained plexus of veins, which constitute the principal outlet. Now in order that the means of escape may be easy, it is above all things necessary, that the stream of fluid through the angle of the anterior chamber should be unimpeded. But this route can be free, only when the angle is widely open.

If it be closed, by the anterior surface of the iris being pressed against the posterior surface of the cornea, the outflow becomes, in great measure, perhaps entirely, arrested. It may now be readily seen how dilatation of the pupil, from whatever cause, may act in the production of an attack of glaucoma, in an eye predisposed to it, or may intensely aggravate an attack already begun. The effect of gathering the iris into folds, towards the angle of the anterior chamber, which takes place when the pupil is fully dilated, is sufficient oftentimes to complete the blocking of the angle, and so to stop the means of exit of fluid, and precipitate the occurrence of all the symptoms which attend its retention. It is always well, therefore, in advanced life, to examine the tension of the eyeball before using atropine, and above all things, not to mistake an attack of glaucoma for an attack of iritis. This mistake will not be made if attention be paid to the following diagnostic points, *which refer to the acute forms.* The chronic forms of glaucoma are not likely to be mistaken for iritis ; there is no congestion of veins, no redness nor pain. In iritis, and in *acute* glaucoma, there is always considerable vascular congestion of the anterior tissues of the eyeball, but in glaucoma the tension is always more markedly increased, the eyeball having a peculiar hard, unyielding feeling, and the pupil being more or less *dilated*, the dilatation being usually, not evenly circular, but oval, with the long axis vertical. Whilst in iritis, the tension is normal, or at most only a little full, never approaching hardness, and the pupil is *markedly contracted*. The pain, too, is somewhat different in character.

In iritis it is generally a burning, throbbing pain in the eyeball, and when it extends to surrounding parts, is generally most marked on the temple and on the inner side of the orbit, extending down by the side of the nose; though it is often severe, it is rarely of extreme severity.

The pain of acute glaucoma is, on the other hand, terribly severe. The eyeball feels as if bursting, and the pain radiates through all the branches of the ophthalmic division of the fifth nerve. Severe headaches, and nausea, and vomiting, and even delirium sometimes accompany it.

In inflammation, therefore, with *contracted* pupil and normal or only slightly increased tension, mydriatics are safe; in cases in which there is doubt as to increase of tension, it is better to run the chances of the possible risks attending their non-employment for a short time, rather than to bring about the positive disaster which their use in glaucoma inevitably entails.

In spasm of the accommodation, the use of mydriatics is distinctly indicated. By this term is meant a tonic spasm, or cramp of the ciliary muscle, such as is sometimes found as the result of injury or irritation, or most usually as the result of excessive, or too prolonged exercise of this muscle in eyes the seat of faults of refraction. We know that patients suffering from hypermetropia, for example, cannot see objects clearly at any distance, either near to or far away, without exercising their accommodation ; that is, without accommodating the curvature of their crystalline lenses to the degree required to focus clear images on their retinæ, by calling into action their ciliary muscles. The constant strain thus kept on the muscles produces a state of permanent tension, which frequently passes into tonic spasm, and at last, the accommodation becomes permanently overstrained, and kept adjusted for vision of near objects only. The optical complications thus induced are themselves a source of serious trouble, but, as is usual in spasm, there are also considerable symptoms of discomfort in the eyeballs, and in the parts about the eyes. To relieve these symptoms, as well as to relieve the optical complications, it is necessary to cause relaxation of the spasm, by paralyzing the ciliary muscle. When this relief is needed it must be fully afforded. To prescribe a weak dose of a mydriatic is to aggravate the evil, such an expedient simply weakens the power of the ciliary muscle, and makes it less able to act, whilst it does not render it incapable of responding to stimulus. Full paralysis of the muscle relieves by rendering response to stimulus impossible.

Mydriatics are of great use in the treatment of phlyctenular and other ulcers of the cornea when they are accompanied by increased vascularity and congestion. They anæsthetize the extremities of the branches of the fifth pair of nerves, and in a marked manner control the superficial vascular supply by their constringent action on the blood-vessels. Ulcers, however, situated at the margin of the cornea, should be treated cautiously by mydriatics, because if perforation should take place, the iris, being drawn towards the margin, is placed in the most favourable position for prolapse and subsequent incarceration in the cicatrix.

In the asthenic forms of corneal ulcer, such as are found in patients whose nutrition has suffered as the result of exhausting diseases, in badly nourished children, and also in women reduced by excessive lactation, together with other weakening causes, mydriatics are not of foremost value, because they tend to produce, at least in the early stages of their application, an anæmia of the structures involved rather than a hyperæmia, and to lessen secretory power, which is obviously not what is needed. For the same reason in those forms of abscess, which occur in course of severe conjunctivitis, or purulent and gonorrhœal ophthalmia, or as the result of direct breaking down and ulceration of centres of infiltration, mydriatics are not indicated.

The therapeutical uses of cocaine call for a more extended notice. This remedy has passed so speedily into the hands of the general body of practitioners that it is now used almost indiscriminately in all cases of eye disease which come under their notice. It is well, therefore, to know the limits, as at present ascertained, within which it may be used wisely and beneficially. As a means of producing anæsthesia of the ocular surface, cocaine is invaluable. If a four per cent. solution be dropped four or five times in course of ten minutes, on to the corneal or conjunctival surface, such a degree of insensibility of those structures will be produced as will permit of any operation, not only cutting operations, but the application of caustics, and of the actual cautery, being performed without pain. In facilitating the extraction of foreign bodies impacted in the corneal surface, cocaine is most useful. It is useful, too, in conditions of corneal disease where spasmodic closure of the lids with corneal irritation, the so-called photophobia,* renders inspection of the eye difficult.

Cocaine is, however, not an unmixed blessing, for it has in some cases a very peculiar and even injurious effect upon the corneal epithelium. After a solution of even five per cent. has been employed, frequently for some time the cornea, especially if it be exposed, shows marked changes. The epithelium becomes dried and damaged, its anterior layers become thinned and flattened, then the deeper layers begin to shrink, and ultimately the external cells are cast off. This is due to diminution of the lymph supply, a

---

* I use the term *so-called photophobia*, because the inability to open the eyelids, which is characteristic of certain forms of keratitis, is not really due to dread of light but to local irritation of the nerve fibrils of the cornea by presence among them of invading leucocytes. It is really a spasmodic closure of the eyelids—blepharospasm—due to the irritation. True photophobia is the result of increased sensitiveness of the retina.

result of the anæsthesia, aided by evaporation. When this condition of disturbance of the superficial layers of the cornea is produced, it may be easily conceived that solutions of otherwise harmless agents may cause troublesome changes. And in diseased conditions of cornea, in which the entrance of micro-organisms into the tissues is a probable complication, it is most necessary to avoid lowering the vitality, and so weakening the resisting power of the cornea, by the persistent maintenance of the anæsthetic effect of strong solutions of cocaine.

In the use of mydriatics there are two or three conditions in which the utmost caution must be observed. They should be carefully watched in the case of old persons and children on account of their toxic effects. These are produced by some considerable quantity of the solution finding its way down the tear passages into the throat. Merely dry throat with some headache and giddiness may follow, or the condition may go on to actual delirium. It is within the experience of those of us who use strong solutions of these drugs, to see old people become what is called "moithering," and even get up and wander about in mild delirium under their influence ; and to see children become restless, feverish, and the subjects of hallucinations. These effects warn us to lessen the frequency of application, if not to stop it altogether.

The disturbing effect of mydriatics upon vision must also not be forgotten. Many a patient has passed hours of misery in fear of blindness through the instillation of atropine into the conjunctival sac, for want of a few words of timely explanation of the effect upon vision.

I am the more inclined to lay stress upon the cautions necessary in the use of mydriatics because it is a matter of observation that in the treatment of diseases of the eye atropine is becoming adopted as a routine expedient. Its unguarded use seems to me to be prompted by some such sage aphorism as that which old players at whist have formulated for the guidance of vacillating neophytes in its mysteries—"When in doubt lead trumps." Now some such loose rule seems to be becoming imported into the general, as distinguished from the special, treatment of eye diseases. "When in doubt use atropine," seems to be the guiding maxim. Against this happy-go-lucky method, I would, for the reasons above given, utter an earnest warning.

MYOTICS.—I shall now turn to the consideration of the actions and uses of myotics, the contractors of the pupil.

Those ordinarily employed in practice are derived from calabar bean and jaborandi, though other drugs, such as opium, morphine, coniine, and digitaline contract the pupil when applied locally.

In practice the most active and powerful myotics are those derived from calabar bean — the extract, and the alkaloid eserine. Formerly the extract was greatly used, but it is so uncertain in strength, and so liable to decompose in solution, that it has been superseded by the alkaloid eserine. The salts of this alkaloid most used are the sulphate and the salicylate.

Many experiments have been performed in order to ascertain the exact action of myotics on the iris. In the case of eserine it is highly probable that it paralyses the peripheral sympathetic nerve fibres, and it is also highly probable that it stimulates the terminations of the branches of the third pair to the iris, so that whilst there is a passive myosis due to paralysis of the sympathetic supply to the iris, there is also an active myosis, due to direct stimulation of the sphincter. The effect is not limited to the iris. Full doses stimulate the ciliary muscle, and so increase the refractive power of the eye. This may be proved by examining eyes in which the iris is wanting, which have been subjected to the influence of strong solutions of eserine. In all cases, however, the effect upon the sphincter of the iris is greater and more lasting than that upon the ciliary muscle.

In testing the comparative duration of the effects of eserine upon the sphincter of the iris, and upon the ciliary muscle in a healthy eye, a recent author found that whilst the effect of a solution of a certain strength upon the ciliary muscle lasted slightly over an hour and a half, the effect upon the sphincter of the iris lasted seventy-two hours.

The pupillary contraction produced by myotics is much

more considerable than that produced by the strongest light. It is rarely equal in each part of the iris, consequently the pupillary opening may look somewhat irregular in shape, more contracted in some parts than in others.

The sulphate of eserine is the salt most generally used. A solution of as much as four grains to the ounce of water may be occasionally employed, but, as a rule, solutions of from one quarter of a grain to two grains to the ounce answer every useful purpose.

Pilocarpine, the active principle of jaborandi (pilocarpus pinnatus), has come considerably into use for the same purposes as eserine. It is best used in the form of nitrate or hydrochlorate of pilocarpine, and its action is to all intents and purposes the same as that of eserine, and therapeutically they may be grouped together. It is, however, a weaker drug than eserine, and needs to be used in a solution at least four times the strength of eserine. It has the advantage of being less irritating locally than eserine, and of remaining longer undecomposed in solution.

Now as to the therapeutical uses of myotics. First of all, they are of signal value in glaucomatous states of the eye. For just as mydriatics are capable of inducing a condition of increased tension of the eyeball, by puckering up and retracting the folds of the iris into the angle of the anterior chamber, and so obstructing the outflow of the intra-ocular fluid by blocking the filtration area ; so on the other hand, myotics relieve the condition of impeded outflow, by contracting the pupil and stretching the iris out of the angle of the anterior chamber, and so widening the interspaces of the ligamentum pectinatum, through which the retained fluid finds escape. As a forerunner of operative treatment, and as affording an immediate means of at least temporary relief in cases of glaucoma, before operation can be resorted to, myotics are of undoubted value.

In the treatment of mydriasis, the result of interference with the action of the branches of the oculo-motor nerve, whether from the result of syphilis or rheumatism, the local stimulant effect of myotics is valuable as an aid to the general and special constitutional treatment.

In corneal ulcers of asthenic type, such as are found in infants and adults suffering from mal-nutrition, and in all cases of corneal abscess and ulcer where necrosis is rapid and perforation threatened, myotics are very valuable local remedies. They produce rapid increase of vascular supply and augment secretion, and at the same time reduce the tension of the eyeball, thus increasing chances of repair and reducing risks of perforation. Their direct value is easily estimated, when we remember that the course of the nutrient stream is from the margin of the cornea towards its centre ; and also that the nutrient supply comes, for the superficial corneal layers, from the conjunctiva, and for the deeper layer from the sclerotic. The beneficial action of a remedy which lessens tension of the eyeball, and so removes interference by pressure, with the currents of the vascular supply and its products, is thus very apparent.

And if an ulcer be threatening perforation, it may be easily understood that the intra-ocular pressure, even if not abnormally increased, acts with relatively greater force upon the thinned and ulcerated part of the cornea, which offers less resistance, than upon the firmer healthy portion.

Diminution of the internal tension by the use of myotics here becomes of primary importance, and should be at once secured.

I will now call your attention for a moment to the interesting physiological fact, that if a mydriatic and a myotic, equal in strength, be introduced into the eye together, the myotic acts first, but is soon overcome by the mydriatic. This points to the practical deduction, that if it be necessary to counteract the effect of a mydriatic, atropine, for example, as in the case of the precipitation of an attack of glaucoma by its incautious use, it is not sufficient to rest contented with one or two applications of a myotic, but to keep up the effect by repeated instillations, so that the atropine may not re-assert its influence.

Now to sum up briefly the indications and contra-indications for the use of mydriatics and myotics in treatment.

First, the indications for the use of mydriatics.

(*a*) Their sedative use, to relieve pain and reflex spasm where the terminations of branches of the fifth pair of nerves are irritated, either by injury, or by the presence of invading leucocytes, as in certain forms of corneal infiltration, especially in phlyctenular keratitis.

(*b*) The dilating power is a most valuable aid to other therapeutical treatment in inflammation of the iris, by drawing it outwards and preventing its adhesion to the capsule of the lens, and also in wounds and in perforating ulcers near the centre of the cornea, by drawing the iris away and preventing its incarceration. It is also useful in permitting an extended examination of diseased and abnormal conditions of the deeper structures of the globe, through the widely opened pupil.

(*c*) Their paralysing power is of use in overcoming spasm of the accommodation, and by suspending the action of the ciliary muscle, to permit an accurate estimate to be gained of the nature and exact amount of refractive errors.

The contra-indications in the case of mydriatics are to be noted :

(*a*) In cases of increased intra-ocular tension except where there is increase of tension with *absolutely entire posterior synechia*, here the well-known action of atropine in limiting secretion may be of use, if operative measures be not warranted.

(*b*) In diseases of the ciliary body, when the eyeball is becoming soft.

The indications for the use of myotics are :

(*a*) In natural and artificial dilatation of the pupil when it is necessary to overcome it.

(*b*) In wounds and perforating ulcers at the corneal margin, in order to draw the iris inwards, and prevent its prolapse and subsequent incarceration in the cicatrix.

(*c*) As a stimulant locally in cases of impaired function of the accommodation, such as the paralysis of the ciliary muscle following diphtheria.

(*d*) To diminish intra-ocular pressure by stretching the iris, and so restoring the routes for passage of fluids through the openings in the ligamentum pectinatum.

(*e*) To influence nutritive supply by lessening the tension in corneal ulcers with suppurative tendencies.

The contra-indications in the case of myotics are :

(*a*) Their prolonged use is liable, through the great vascularity they encourage, to bring on a low form of iritis, and so give rise to adhesions between the iris and the lens capsule.

(*b*) They should not be used in secondary glaucoma with complete posterior synechia, because of their power of increasing the intra-ocular secretion.

(*c*) Toxic effects should be watched for, and when present should cause the administration of the drug to be immediately suspended.

SECTION II.

ASTRINGENTS, STIMULANTS, AND IRRITANTS.

Astringents, stimulants, and irritants are local remedies which may be appropriately classed together, because the differences in their actions are for the most part differences of degree rather than of kind. They vary in intensity of effect rather than in diversity of method. Thus, the same substance used in solutions of varying strengths may be either astringent, stimulant, or irritant.

When the slightest, or purely astringent effect is desired, it is in order to bring about a simple contraction of the tissues, and so to restore a relaxed tissue to its normal degree of firmness. This is necessary in certain stages of inflammation, and it is probably limited to producing a more or less definite degree of contraction and condensation of the relaxed tissues, with which the reagents employed are brought directly in contact.

If the same reagents be used in stronger solutions than are needed to produce the astringent effect, they become stimulants, and by restoring the tone of the blood-vessels and rousing the natural forces, they stir up flagging curative action, either by hurrying on repair of lost tissues, or by promoting the regression and absorption of inflammatory products, and so restore the weakened and altered tissues to their normal conditions.

Still stronger applications of the same substances produce irritant effects, and give rise to dilatation of the vessels. This dilatation may be only of sufficient extent to produce intense redness and congestion ; or, if pushed still further, it may produce an actually disorganising inflammation.

The slighter effects of these remedies are useful in the simpler forms of conjunctival inflammation, and in the earliest and latest stages of the more severe forms ; and in the relaxation which follows the over-excitement of inflammation generally. Their more profound effects are reserved for the more active and severe stages of inflammation of the conjunctiva, when speedy changes in nutrition, or alterative action in diseased tissues and their secretions, are needed.

I will take first of all the astringent action, and the special indications for its employment.

In the conjunctiva, as in all tissues, there is a normal degree of compactness, which, during the progress of an inflammation, becomes loosened by the engorgement of the vessels and by consequent exudation of plasma and invasion of leucocytes. And, like inflamed mucous membranes generally, the conjunctiva undergoes great relaxation, owing to its lax structure and yielding attachment to the firmer tissues beneath it. When the acuteness of the inflammation is past, the condition of relaxation remains, accompanied by more or less muco-purulent secretion from its surface. The restoration of tone and the consequent checking of secretion, are the objects sought by the use of pure astringents, or of more powerful re-agents, in the limited astringent degree of their action.

After care has been taken to distinguish and remove as far as possible the exciting cause, this astringent action is appropriate in the relaxed conditions of the conjunctiva which follow the excessive hyperæmia induced by cold and the irritation of foreign bodies, affections of the lachrymal passages, exposure to irritating vapours, and the effect of over-strain of the eyes. It is equally appropriate in the chronic stages of follicular conjunctivitis, and in the initial and terminal stages of the more severe forms of conjunctivitis —catarrhal, blenorrhœal, and purulent.

The astringents in general use are the pure vegetable astringent tannin, which in the form of the glycerine of tannin, or still better, of an ointment compounded with vaseline, is very useful in simple relaxations of the conjunctiva, and in the later lingering stages of follicular and granular conjunctivitis. The mineral astringents in general use are the nitrate of silver in the proportion of half a grain, to a grain, to an ounce of water ; perchloride of mercury in the proportion of one-eighth of a grain ; the sulphate and chloride of zinc, the sulphate of copper, alum, and the acetate of lead—in the proportion of from half a grain to two grains to the ounce of water.

Closely allied to the astringent action of these remedies, apparently only an extension of it, is their stimulant action.

By using them in increased quantities, that is to say, in the proportion of four or five grains of the sulphate of zinc, or alum ; two grains of the chloride of zinc, or sulphate of copper ; two to four grains of the nitrate of silver ; or a quarter to half a grain of the perchloride of mercury—all to the ounce of water, full stimulant effects are produced. In this category should be included one of the favourite local eye applications of the general profession, the vinum opii, which owes its value to the alcohol in its composition more than to the opium. It is little used—if at all—by ophthalmic surgeons.

These stimulant effects are useful in the various forms of conjunctivitis, when the acuteness of the inflammation has in a measure subsided, and the bright red colour has given place to a more livid hue. Or, as before mentioned, when in the later stages, repair is at a standstill, or inflammatory products are slow in being absorbed.

Besides the stimulants just mentioned, which are used in solution, there are others used in the form of dry powder, or made up into ointments.

The stimulant powders in general use are calomel, and boric acid. These are dusted into the conjunctival sac, and are employed in the later stages of interstitial keratitis, when the infiltration is slow to clear away ; and in phlyctenular keratitis, when irritation and the so-called photophobia have subsided. They are also useful in phlyctenular conjunctivitis, and in indolent and vascular ulcers of the cornea.

Boric acid is a harmless substance, and produces no ill effects ; but calomel, unless carefully watched, may produce much mischief. If a considerable quantity of calomel be dusted into the conjunctiva, some is apt to remain in the lower palpebral fold, and collect into an irritating mass. Whether this is due to the chloride of sodium in the tears acting upon it and converting some of it into an irritating perchloride of mercury, as is mentioned in the text-books, or not, I am not prepared to say. But I have more than once seen a large superficial slough of the lower palpebral conjunctiva follow such a careless application.

Stimulating ointments act very powerfully, and are most useful. The salts of mercury, especially the yellow oxide, the red oxide, and the nitrate, all in free dilution, act very beneficially in chronic conjunctival diseases, and in the various forms of keratitis, when the acute stage is past, and the infiltrations remain unabsorbed. The ointments of the yellow and red oxides are best used in the strength of half a grain, or a grain, to the drachm of benzoated lard, or vaseline ; and that of the nitrate in the proportion of five to ten grains of Ung: Hydrarg: Nitrat: to the drachm of lard or vaseline. Besides being useful in corneal and conjunctival affections, these ointments are most beneficial in blepharitis, by which is meant the inflamed condition of the muco-cutaneous margin of the eyelid, so common in badly-nourished children, especially among the poor.

The irritant effect of local applications is really only an aggravation of the stimulant effect, and it is but rarely that it is needed. It is only wanted when the response to stimulus is very slow and imperfect, and care must always be taken that the effect be not carried too far, and actually troublesome inflammation set up. So far as their use in the treatment of superficial inflammations of the eye and their results are concerned, it is sufficient to direct to your attention what I have said as to the use of stimulants.

There are two especial cautions to be observed in the use of this class of remedies. The first refers to the whole of them collectively, and emphasizes the injunction, that their use is strictly limited to superficial affections of the eye. Ignorance or neglect of this simple caution has led to the causation of much needless suffering, and to irreparable damage to sight. Irritating remedies should never be used when there is any evidence of disease of the deeper membranes of the eye. It is by no means uncommon for ophthalmic surgeons to see cases of sclerotitis, and iritis, and even choroiditis, greatly aggravated by the use of local irritants-through want of knowledge of the true character of the disease. In all these affections there is more or less redness and congestion of the superficial vessels of the eye, but it must not be supposed on that account alone that astringents or stimulants are indicated. An attempt, at least, at more accurate diagnosis, should be made before deciding on local treatment. In order that a mistake may not be made at the outset of an inflammatory attack, as to whether it is a conjunctivitis, or an iritis in the earlier stages, care should be taken to notice the action of the pupil in the rapidity and regularity of its response to light, and to judge by comparison of the irides of the two eyes, whether the iris of the inflamed eye has lost any of its brilliancy of colour. The character and degree of the pain attending the attack should also be noted, bearing in mind that the pain in inflammation of the deeper membranes is in excess of the apparent inflammation, whilst in superficial inflammation the redness and congestion are altogether out of proportion to the pain experienced. And in distinguishing between sclerotitis, with or without an early iritis, there is one simple means of differentiation, which is not generally taught, and which the most inexpert may employ. It consists in taking the lower eyelid under the point of the forefinger and using it to sweep the surface of the eyeball from the cornea downwards, making at the same time firm pressure on the globe. The bloodvessels of the ocular conjunctiva will be for the moment emptied, and the white sclerotic will be seen clearly, if the vascular engorgement be limited to the conjunctiva. But if any congestion of the sclerotic, such as accompanies sclerotitis, and iritis, exist, the well-known peri-corneal pink zone of straight vessels will be uncovered, and will remain visible and unaltered by the pressure. Whenever this peri-corneal pink zone is present, no local irritant must be used.

The second important caution refers to one mineral astringent, lead, and its salts, which I have refrained from noticing specially earlier in these remarks, in order that my notice of them may receive greater emphasis—I allude to the acetate of lead in particular. This salt has become a stock, routine remedy in most dispensaries in the treatment of superficial inflammations of the eye, and lead lotion seems to be prescribed almost as a matter of course by many practitioners. Against this custom I would express an earnest protest. Acetate of lead has a dangerous peculiarity, which should exclude it from the list of local applications to the eye. It has no advantage over other remedies of this class, and it has the enormous disadvantage, that if it is brought in contact with a cornea having its epithelium ever so slightly abraded, it at once enters into chemical union with the corneal tissue, and forms a dense, opaque, permanent film. Many eyes have been irreparably damaged by the use of acetate of lead lotion, which, without it, would probably have made a sound recovery.

SECTION III.

CAUSTICS.

I come now to touch on caustics and their uses in the distinctly purulent forms of conjunctivitis. In these cases the salts of silver and mercury are the most trustworthy, and among them all the nitrate of silver stands foremost in value. In all cases of catarrhal, blennorhœal, and purulent ophthalmia, from whatever cause, the nitrate of silver in proportion commensurate with the stage of activity of the disease is the surest remedy. In the earlier stages, whilst the discharges are watery and mucous, it must be used in weak solution, one or two grains to the ounce of water; in the more active stages, where copious purulent discharge is present, it is necessary to use it from four to thirty grains to the ounce. But it must be borne in mind in using strong applications of the nitrate of silver, or any other caustics, that it is above all things necessary to limit their action to the parts actually needing it. To ensure this the caustic should be applied carefully, and all superfluous particles should at once be neutralized and washed away by copious affusions of warm water. In the case of nitrate of silver the best neutralizing agent is chloride of sodium; contact with this converts all the superfluous nitrate of silver into the chloride of silver, which is inert.

In very severe cases of purulent ophthalmia, especially

those owning a gonorrhœal origin, the nitrate of silver may, by those accustomed to it, be used in the form of mitigated stick ; that is in the form of one part of nitrate of silver, melted with two or three of nitrate of potash, and moulded into shape. When this solid form is used to touch the conjunctiva great care must be taken to keep the cornea free, and to follow the application immediately by free washing with a strong solution of chloride of sodium.

The perchloride of mercury is useful in the same classes of cases as the nitrate of silver, but the irritation following its use in strong solutions is greater and more prolonged than that following the nitrate. For its caustic effect a solution of one part to four hundred is sufficient. Like the other caustic solutions it should be brushed over the parts affected and quickly followed by free washing with warm water.

In the use of all caustic applications to the conjunctiva a very simple rule should always be borne in mind for guidance, viz. : the more profuse the purulent discharge the stronger the solution required ; the more watery the discharge or the drier and more infiltrated the conjunctiva the weaker the solution required.

SECTION IV.

ANTISEPTICS.

Many of the remedies classed as stimulants and caustics possess decided antiseptic qualities, and in all those forms of conjunctivitis, which are due to the intrusion of micro-organisms, the antiseptic action becomes the most important. It is well known that the simpler catarrhal or muco-purulent form of conjunctivitis when intensified by rapid transmission from subject to subject, living under bad sanitary conditions, may become so intensely purulent as to be indistinguishable clinically from forms of more directly specific origin. Thus in one case of infection from a purulent source (and I believe all purulent discharges to be more or less infective), we may see a simple muco-purulent conjunctivitis ; and in a second the development from the same source of infection may present an acute purulent form. This we may attempt to explain by supposing a difference of soil, and a varying yield, though the germ may be the same. Certain it is that identical pathological products do not always, when brought in contact with corresponding parts in different persons, produce exactly similar results. In the discharge in all these forms of purulent ophthalmia there are found as the basis of the mischief various micrococci, and the antiseptic action of the remedies of which I am about to speak is directed to their destruction.

In the earlier stages of the infection when the conjunctiva is but slightly congested and the discharge is watery or muco-purulent, free and persistent irrigations with weak solutions are more useful. For example, of boric acid 1 to 50, salicylic acid 1 to 200, carbolic acid 1 to 300, aqua chlori, nitrate of silver ½ to 100, hydrarg : perchlor: 1 to 5000. Later on, when the discharge is thoroughly purulent and fully established, the nitrate of silver 3 to 6 per cent. and the perchloride of mercury 1 to 500 or 600 are alone to be depended upon, and they must be applied in the careful manner, and superfluous portions should be neutralized, as described when treating of their caustic action.

In addition to these remedies iodoform is spoken of by some authorities as of substantial value in the treatment of these septic forms of conjunctivitis, but it has really little or no value as a germicide, and I would warn you against trusting to it alone.

Resorcin and its ally hydroquinone are very useful in 1 per cent. solution, and I have used iodol in form of ointment 1 to 10 with distinct advantage in conjunctivitis, and in infective corneal ulcer.

The yellow oxide of mercury and the weak nitrate of mercury ointments, in the strengths before noted, are very useful as lubricants and germicides, and form valuable adjuncts in treatment.

As a practical instance of the method of use of

antiseptics I think I cannot do better than just briefly sketch the guiding points in the treatment of that disastrous form of contagious ophthalmia known as ophthalmia neonatorum.

When on the second or third day after the birth of the child its eyelids become a little swollen and red at the margins, and the conjunctiva congested and a slight muco-purulent discharge begins to show itself, the child's head should be laid well back and the eyelids widely opened and the conjunctival sac thoroughly washed out by means of a stream of the following lotion, squeezed from a saturated mass of cotton-wool held a few inches above the eye—acid boric 1 to 50 with perchloride of mercury 1 to 4000. This should be done at least every hour, and a weak yellow oxide of mercury, 1 to 120, smeared over the edges of the lids. Later when, if not checked, the discharge becomes purulent and the swelling and congestion of the conjunctiva increase, the eyelids should be turned outwards so as to throw forward as much as possible the deeper conjunctival folds, and the whole conjunctival surface should be painted with a solution of nitrate of silver, grains 15 to water one ounce, by means of a camel hair-brush, sweeping it repeatedly into the upper and lower culs-de-sac, and the superfluous nitrate should then be neutralised by a solution of chloride of sodium. These methods of the application of antiseptics should be kept up as the groundwork of the treatment during the whole course of the attack, and they serve to stand as types for the use of antiseptics in the other septic forms, the only variations being in degree of activity or lenity according to the stage or the character of the disease.

In the operative surgery of the eye the employment of antiseptics has undoubtedly contributed to a greater proportion of successful results. Whether this is due purely to the antiseptics used or to the cleanliness and care which has followed in their train I am not now going to discuss. But I have no doubt of the truth of the general proposition that the use of weak solutions of the perchloride of mercury and boric acid to cleanse the conjunctival sac before operating for cataract, has been the means of saving many eyes which without this precaution would have been lost from suppuration.

Chapter II.

HEAT AND COLD.

AMONG the means of controlling febrile action and limiting local changes in diseases of the eye, heat and cold occupy important places. The addition of heat and its abstraction have each a considerable value in ophthalmic therapeutics.

The effects of the application of heat differ somewhat in detail according as it is applied in a dry or moist form.

Dry heat excites and stimulates the nervous and vascular supply of the part to which it is applied. It causes local redness and hyperæmia with increase of temperature ; in fact its action is directly stimulant.

Moist heat on the other hand, besides stimulating the vascular supply, has a softening and relaxing effect upon the parts to which it is applied. It lessens tension and relieves the pain which accompanies the fulness and engorgement of the vessels in acute inflammation.

Dry heat is therefore useful in a different class of cases to those which are benefitted by moist heat.

In cases in which the vitality of a part is much lowered, such as in corneal ulcers in which rapid necrosis of corneal tissue is going on, dry heat applied by means of heated cotton wool, in masses frequently changed, is most useful ; and steamed pads, consisting of several thicknesses of lint heated over steam and squeezed perfectly dry and applied as hot as can be borne, stimulate the vital action of the parts and rapidly bring them up to normal conditions.   The action of dry heat in sloughing corneal ulcer is greatly assisted by the instillation of eserine solution gr. 1 to water one ounce every four hours.

Moist heat in the form of hot pads of lint covered with oiled silk, and as fomentations, is useful in superficial inflammations which have existed for some time, and in which there is much tension of the parts inflamed.

A moderate warmth is sedative in character, and in the early stages of superficial inflammation acts in some measure in checking their progress, but in these stages it is less reliable and effective than cold.   When inflammation has gone on to suppuration, moderate heat is of decided advantage by its relaxing influence, in furthering the formation of pus and its discharge, and so hastening the progress of the case.

These milder degrees of moist heat are of use chiefly in superficial inflammations.   In deeper seated inflammations, such as those of the iris and sclera, a greater degree and a more persistent application of heat is useful.   The effect of

such heat, which is best applied by means of hot fomentations, is to influence the circulation in these tissues in a
marked degree. The case afforded in iritis and sclerotitis
by hot (not merely warm) fomentations, vigorously and
unremittingly applied, is beyond all question.

Excessive heat it must however be borne in mind is a
depressant, and, if very intense, causes decomposition of
tissue.

The application of cold has a wider field, and, unlike
moderate heat (by the use of which it is exceedingly
difficult, if not impossible, to do harm), it requires careful
watching in its application.

Cold continuously applied lowers temperature and
diminishes sensibility, and it causes contraction of vessels
and tissues and so reduces the size of the part to which it
is applied.

The application of cold for short periods, with intervals,
is followed by considerable reaction, and heat returns
perhaps in excess of the degree originally present.

The abstraction of heat, which the continuous application of cold involves, lowers the vital activity of the
part, and the smaller arteries and capillaries become contracted. In this way the quantity of blood carried to the
parts influenced becomes lessened, and the ordinary physiological and chemical processes required for the maintenance
of vitality of the tissues are interfered with, and functional
activity is diminished. This is, as a rule, the utmost effect

required to be produced by the application of cold in surgery. Iced applications kept up for a considerable time have been known to cause gangrene.

Cold, therefore, whilst a very valuable therapeutic aid, is indicated to a limited extent only in inflammations of the eye. When it is so employed it should be continuously applied until its effects are satisfactorily produced, and then discontinued ; if remittingly used, contrasts of temperature are produced, which are not beneficial.

The value of cold when applied to the eye is observed especially in traumatic conditions—wounds and contusions of the eye—and in attacks of glaucoma where the patient will not submit to operation. Combined with the frequent use of eserine, an acute attack of glaucoma may be sometimes overcome, doubtless by the effect of the cold on the swollen iris and ciliary processes. The shrinking produced in these parts opens the meshes of the ligamentum pectinatum, previously closed by the pressure, thus permitting increased outflow of the retained fluids.

The most frequent method of applying cold is by means of compresses of iced water, kept cool by contact with ice and squeezed fairly dry.

The objection to these compresses is, however, very considerable on account of the dripping wetness caused.

A better plan is to enclose the ice in thin india-rubber, or a child's rubber balloon, which by their thinness allow of full effect of cold and at the same time are quite dry.

The receptacle may be filled with finely pounded ice and suspended round the forehead so as to press but little on the eye.

The tubes known as Leiter's tubes, in which a small fine coil is laid over the eye and a slow stream of cold water kept running through it, are good, but their weight is objectionable.

There can be no doubt that wounds and contusions of the eye recover more quickly under the application of cold, with less pain and disturbance, than under any other treatment. Supposing, of course, that the patient is possessed of fairly average vitality ; if he should be of very cachectic habit, heat is likely to be more useful.

At the *outset* of all forms of inflammation of the superficial structures of the eye, conjunctiva, cornea, and even of the iris, cold is often most useful.

There are, however, some patients who either from some peculiarity of nervous constitution or other obscure cause experience shows to be unable to tolerate cold as an application. In such persons heat, even in the acute stage of inflammation, gives greater relief, the relaxing effect of warmth appearing more grateful and beneficial than the contracting effect of cold on the circulation.

SECTION II.

BLOODLETTING.

I come now to consider the methods and value of blood-letting in the treatment of eye diseases. It is, it may be said at once, unnecessary to discuss the value of venesection. To be of any use in affecting the circulation of the eyeball venesection must be carried to a very considerable extent, and any benefit which might be gained would be more than counterbalanced by the injurious effects produced upon the general bodily nutrition. It is therefore to the local abstraction of blood that I shall confine your attention.

Formerly, and still by some practitioners, local abstraction of blood was considered the chief, if not the only, trustworthy means of combating hyperæmia and relieving pain in inflammations of the ocular structures. When rightly estimated, however, the effects of this procedure upon the local circulation are found to be so transient, that it can only be regarded as of temporary value, and an adjunct to other means more lasting and settled in their results.

In all cases of inflammation there are exacerbations and remissions, and at the period of exacerbation the hyperæmia is greatest, the inflammatory products most abundant, and the pain most severe. Employed at the period of exacerbation in ocular inflammations local blood-letting reduces the activity of these processes and gives temporary relief, and by a slight emptying and reduction of tension of the overgorged vessels it increases the chances of absorption of remedies locally applied. It also gives the turgid vessels, which have lost their contractile power from continued over distension, an opportunity of recovering their tone, by lessening the quantity of their contents. But this is all that can be reasonably expected to be gained by local bloodletting. It can only be looked upon as a physical ally in our endeavours to relieve pressing symptoms. The congestion and pain are relieved, but with only slight, if any, enduring effect upon the course of the disease. For it must be ever remembered that withdrawal of blood from an inflamed part is inevitably followed by reaction, and, if the hyperæmia, which it is intended to relieve, occur in a strong subject, the reactionary afflux of blood may be so considerable as to place him in a condition as bad, or possibly even worse, than before. The necessity for control of the reaction after bloodletting must therefore be constantly borne in mind. Another point, too, must be borne in mind in assessing the value of local bloodletting, and that is, that the most satisfactory effects are obtained in superficial inflammations ; in the deeper-seated forms, where its action would be more valuable, it is found to fall greatly short of what is desired. This is probably due in great measure to the anatomical relations of the parts involved. The connections between the blood-vessels of the temporal and mastoid regions, and those of the interior of the eyeball, being far from intimate.

Local bloodletting may be carried out in two ways. First by scarifications of the conjunctiva. This is done by means of incisions of varying number and depth according to the severity of the case, and is employed in engorgement of the palpebral and ocular conjunctiva.

Second by leeching, either by the use of the natural leech or by the artificial leech which is an improved modification of cupping.

The natural leech is best used where the inflammation is superficial ; for example, in cases of the external ophthalmiæ occurring in full-blooded persons, and attended with great vascular engorgement.

The artificial leech is most useful where active depletion and the reflex nervous effect of the strong suction is required ; for example, in iritis, irido - choroiditis, choroido-retinitis, and inflammations of the deeper structures generally.

Leeching is best applied over the temporal region except where there is reason to believe that the congestion in inflammation is associated with hyperæmia of the cerebral meninges—in such cases the mastoid region should be chosen.

If a more rapid effect than is ordinarily obtained by the natural leech be thought desirable, it may be secured by

practising upon the leech the little operation called
bdellatomy, which consists in making an incision into the
side of the leech whilst it is sucking. The blood sucked
by the leech escapes from the wound in its side, and in this
way one leech will draw an ounce or more.

The artificial leech invented by Heurteloup is the most
trustworthy means of local bloodletting in eye cases. This
instrument consists of a circular knife which is made to
press on the skin and caused to revolve rapidly by means
of a simple mechanism. By this means a circular incision
of the depth desired is obtained, and over this incision an
exhausting glass is placed which is fitted with an air-tight
piston, worked by a fine screw. By arranging the movement
of the screw, suction is made as rapidly as may be desired.

Whenever bloodletting is deemed advisable the time
chosen for it should be that of the exacerbation of the
inflammation, and to ensure its full effect it should be
followed by every means possible to lessen reaction. Among
these the most important are rest in the recumbent posture,
quiet, exclusion of light, and the internal administration of
sedatives and depressants, e.g.: morphia, chloral, aconite,
the bromides. By these means the duration and activity
of reaction are restricted as far as possible.

To the intelligent exhibition of remedies of these classes
at the outset of inflammatory attacks must undoubtedly be
attributed the decline in frequency and extent of bloodletting, which marks the modern treatment of eye diseases.

SECTION III.

COUNTER-IRRITATION.

It has become so much a matter of routine in general
practice to employ counter-irritants in the treatment of all
stages of the inflammations of the various structures of the
eyeball, that to declare the custom to be of doubtful value,
inasmuch as the effects desired cannot be assured with
any approach to accuracy either in degree or in duration,
may seem a startling assertion to many. Yet I have no
hesitation in saying that unbiassed examination leads me
to the belief that the claims of counter-irritation, as it is at
present understood, to a place as a demonstrably serviceable
means of treatment in eye diseases, can receive only largely
conditional acceptance.

Facts observed on all sides show that both in health and
disease an organ may be affected by irritation of a distant
part. The relief to general gouty symptoms, which follows
an acute outbreak in the toe, for example ; and the duodenal
ulcer which follows the irritation of burns of the surface of
the body, are instances of this sequence known to us all.
And, moreover, clinical experience shows that internal
morbid processes are influenced by the production of
external irritations. But between the acceptance of these
general propositions, and that of the trustworthiness of any
existing guide to the employment of counter-irritation in

eye disease, there is a wide, and to me as yet, an impassable gulf fixed.

It is a fact beyond dispute that during the last two generations or so the results of the treatment adopted in eye diseases have vastly improved. It is no less certain that moxas, issues, and setons have almost completely disappeared from the routine of even the most stagnant survivals of oldfashioned practice. Blisters and rubifacients are now the forms relied upon, and even these with a desire for more certain indications as to their use, and with a predilection for their replacement by agents less objectionable in their methods of action. And the explanation of this altered bearing lies not in the fact that the more violent measures are rendered unnecessary by change in type of disease, or by reduced stamina of subject, but in that a truer knowledge of the pathology of eye diseases is gaining ground, together with a more acute clinical appreciation of the value and actions of remedies, with the result that uncertain means of relief are rejected, and more accurate and trustworthy ones are employed to replace them.

Whatever may be the value of counter-irritation in inflammations of the uveal tract, and in congestions of the deeper membranes generally (and I regard it at present as an uncertain quantity), there is one form of disease in which I believe it to be unnecessarily severe and unwarranted by successful results, and that is in phlyctenular keratitis, the strumous ophthalmia of the older writers. For a child already suffering miseries from a phlyctenular keratitis and its ensuing blepharospasm, to be further tormented with a troublesome skin irritation in the form of blisters or setons to the temple, or the nitrate of silver or iodine over the eyebrows and upper eyelids, is to my mind utterly unwarrantable. Yet this is still one of the routine methods adopted in the treatment of this disease, with a view to relief by revulsive action. I would say, when tempted to use counter-irritation in these cases, examine well the bodily condition and remedy any derangements of system, adopt sound hygienic measures, and above all select judiciously and apply perseveringly local means of relief, such as atropine, cocain, and the ointment of the yellow oxide of mercury, and counter-irritation may advantageously be left out of the treatment.

As to the remedies which have in great measure replaced the counter-irritants in the treatment of diseases of the eye, the principal are the local sedatives and anæsthetics, atropine, cocain, and the other mydriatics, and aconite ; and the internal remedies opium and its derivatives, chloral and its derivatives, belladona, hyoscyamus, gelseminum, the bromides, and the salts of mercury and iodine. Of the particular indications for the employment of these I shall speak later.

Finally it is well to bear in mind that counter-irritation is painful in itself, and oftentimes the source of secondary troubles of considerable extent—I mean the eczemas and suppurations and local disturbances which follow it. Bearing therefore in mind its uncertainty, its unpleasantness, and the fact of its replacement in the main by more certain means, I think you will find that counter-irritation may safely be left out of your ordinary calculations and reserved for the few cases which, for want of perfect knowledge of all the factors, seem to resist all ordinary means of treatment. And this may be done not only with no disadvantage to the successful issue of the case, but with infinite gain to the comfort of the patient.

Chapter III.

NARCOTICS.—Narcotics are of great use in ophthalmic medicine, not only by deadening the sense of pain, but also by subduing the bodily and mental restlessness and anxiety which the pain produces. The soothing effects of narcotics upon the sensory nerves are most valuable for their influence upon the circulation of the eye, and secondarily upon its nutrition. This is most marked, for example, in certain acute forms of corneal ulcer, where the pain is extraordinarily severe, altogether out of proportion to the extent of the lesion. These extremely painful ulcers occur very frequently in persons in reduced health, not uncommonly in women exhausted by over lactation, and in old and feeble people as the result of some trifling injury. In these painful forms of ulcer, besides the use, as before recommended, of local antiseptics and sedatives with myotics or mydriatics, opium or morphia, in sufficient doses to deaden the sense of pain, form most valuable items in the treatment. The nervous irritation being subdued, the progress towards recovery is immensely aided.

In the ulceration of phlyctenular keratitis in children
the administration of opium is advisable in carefully adjusted doses. In iritis narcotics afford a very valuable
resource. I have seen cases of plastic iritis treated suc-
cessfully and relieved entirely in a very short time, being
caught in the earliest stages, by the free use of atropine
locally and the steady administration of morphia in doses
just sufficient to keep the patient's sensations blunted, and
this without any attention to the causation of the iritis or to
the internal administration of any specific remedy.

The narcotics relied upon in ophthalmic practice are
opium and its alkaloids. By far the most valuable for
simple relief of pain is morphia. In old persons and in
children it is, I need hardly say, necessary to watch its
administration closely. The use of chloral as a narcotic
after operations in elderly people and in traumatic conditions generally is advantageous, and when combined with
the bromide of potassium it seems in full-blooded persons to
relieve congestion to some extent as well as to deaden pain.

In calming the irritation which so often accompanies
phlyctenular keratitis in children, I have found small doses
of the bromide of potassium and chloral, repeated at regular
intervals, useful. It is certainly as useful in the majority of
cases as the treatment by small doses of antimony with
opium, which is relied upon by some practitioners.

In some obstinate cases of iritis, where relapses of pain
are frequent, in addition to surgical measures such as

I

paracentesis of the anterior chamber, the administration of a mixture containing chloral and morphia gives great relief.

In all cases of eye disease, however, in which a narcotic is required, and the age of the patient permits, morphia either hypodermically or by the mouth is the surest and most trustworthy drug.

ACONITE is a useful remedy in inflammations affecting the fibrous tunic of the eyeball especially—that is to say those in which the sclera is most actively attacked. Thus, in a pure sclerotitis, though such a thing is rarely seen, and in the common sclero-conjunctivitis, and in the form of sclero-iritis in which the sclerotitis is the most marked symptom, aconite is very useful. These affections occur most commonly as the result of exposure to cold. If in an attack of sclero-iritis the iritis should become marked and plastic exudation render the use of mercury necessary, the two remedies, aconite and mercury, may be used in conjunction with the greatest benefit.

In these cases there is usually marked neuralgic pain in the eye and the parts adjacent. This pain yields more quickly to aconite than to any other remedy.

In cases of neuralgia of the eyeball and the adjacent parts, without inflammatory symptoms, aconite is useful.

I think it possible, too, that aconite, from its wellknown paralyzing effect upon the peripheral terminations of sensory and motor nerves, and its therapeutical action in relieving over-excitation of these nerve endings, may be useful in the neurotic disturbances which precede, if they do not absolutely form the foundation of certain forms of glaucoma.

GELSEMINUM is a useful remedy for the forms of neuralgia, in which, besides the eyeball itself, the parts below the eye, supplied by the branches of the fifth pair of nerves, are affected. In a troublesome form of neuralgia in which that branch of the nerve which runs down by the side of the nose is especially involved, gelseminum seems also occasionally useful.

CROTON CHLORAL HYDRATE, or as it is more properly termed, butyl chloral hydrate, is very useful in neuralgic pain radiating from the eyeball, given in ten grain doses every half-hour for three or four hours, or until relief is afforded ; it will oftentimes be found useful when other remedies have failed.

QUININE is a remedy which produces a marked effect upon the eye. There can be no doubt that quinine in very large doses has produced distinct amblyopia—that is to say diminution or loss of vision without any evidence of lesion of structure to account for it. Great diminution of vision has persisted for long periods after the use of quinine in large doses ; one case is recorded (Brit. Med. Jour. 1886.i.823) where vision only began to return after twenty-one days. In some cases a peculiar retinal pallor has been noticed. Von Graefe records two cases of quinine amblyopia, which came under his observation, in one of which a patient had taken half drachm doses till he had taken six drachms ; in the other an ounce of the drug had been taken.

In neuralgia accompanying corneal ulcers quinine is of immense value, and in supra-orbital neuralgia it is almost specific. And in intermittent attacks of pain, in and around the eye, from whatever cause, quinine is most useful.

ALTERATIVES, and especially the salts of MERCURY, are largely used and of prominent value in the treatment of eye disease. If I were to be asked what internal remedy I consider of most value in eye diseases, I should answer that mercury is incomparably the most generally useful and the most trustworthy.

In the diseases affecting the uveal tract—iritis, cyclitis, choroiditis—in optic neuritis ; in the paralysis of the ocular muscles, which is the result of pressure of exudation on the nerves supplying them ; in all conditions where there is inflammation with or without the exudation of plastic material, mercury stands out as the one remedy *par excellence.*

In fact, as soon as inflammation has attacked any of the internal structures of the eyeball, at that moment the indication for the administration of mercury is manifest.

The only question which need exercise the mind of the practitioner is not what remedy shall be administered, but in what form the one appropriate remedy shall be exhibited.

In iritis it is not necessary to inquire into the so-called diathetic influences on which so much stress is laid by most writers—whether the constitutional cause be considered to be syphilis, or rheumatism, or gout—the facts to be really taken into account are the state of the tissues and the degree and character of the local manifestations of disease. If there be iritis with exudation and commencing synechia, that is if the pupil, when submitted to the action of atropine, remains contracted or dilates irregularly, mercury must be given without delay. The object to be secured is the absorption of the lymph and the consequent freeing of the adhesions.

The notion generally held in the profession is that iritis occurs as a result of, or as one of, a series of pathological processes in the course of the evolution of the phenomena of certain diseases, and so it has come to be definitely classed as syphilitic, rheumatic, gouty, malarial, &c., and when a label cannot conveniently be found, as idiopathic.

But if the grounds for this classification of iritis come to be investigated, it will be found that the only warrant for it depends on the coincidence that in certain persons suffering from various diseases iritis has been found to occur.

There can be no doubt that iritis is frequently seen in patients suffering from syphilis, and that it occurs at varying and uncertain stages of the syphilitic process. But there are really no local symptoms, and no combination of them, which allow us to diagnose a given case of iritis as syphilitic. And, on the other hand, if the test of the success of specific treatment be appealed to, it is well-

known that any iritis, whatever may have been its origin or
its peculiar characteristic, may yield to anti-syphilitic treatment. The great danger in thus attempting to label iritis
seems to me to be lest the local manifestations should be
lost sight of in the haze of the diathetic misnomer.

The question of treatment, therefore, resolves itself
primarily into the means to be adopted to recover and
maintain patency of the pupil.

In all the forms of disease in which the exudation of
lymph is the dangerous symptom, the indication given is
to procure as rapidly as possible the therapeutic effects of
mercury.

I find that in most cases the plan of giving grey powder
or calomel in small doses, very frequently repeated, is the
most useful. I give one-eighth of a grain of calomel, with
an equal quantity of opium, or two grains and a half of
grey powder with an eighth of a grain of opium, every two
hours. By this means the degree of constitutional affection,
which it is desirable to reach—namely, the production of
a slight but well marked spongy line on the gums—is quickly
secured. This effect once obtained should be maintained
for some days, or as long as may be necessary. In iritis it
must be kept up until the effused lymph is softened and
absorbed, so that the adhesions may be broken down by
the action of the mydriatic employed. When the deeper
portions of the uveal tract, the ciliary body, and the choroid
are effected, the administration of mercury must be still

further continued, perhaps for weeks, or even months ; and, in these cases, some of the other salts of mercury, notably the perchloride and the iodides, present advantages over calomel.

In choroiditis, and in the form of irido-choroiditis, called sympathetic, that is the form of inflammation of the uveal tract, and especially of the ciliary region, which comes on in the sound eye after injury or disease in its fellow (after the exciting eye has been removed), mercury must be given very patiently and persistently for a long time. The best form is that of the perchloride in small doses.

As to the success of this treatment for sympathetic ophthalmia I fear not much can be said, but my defence of it is that it is the best at present before us. I think I have seen the plastic deposit lessen under its use, certainly, during the time of its administration, and I shall therefore continue to prescribe it in such cases as come under my care.

In the plastic changes which from time to time follow wounds of the eyeball and operations, especially those for the extraction of cataract, evidences are most frequently seen in the ciliary region ; here the frequent administration of small doses of the perchloride of mercury, say twenty minims of the liquor, in some suitable vehicle every two hours, certainly seems to control the exudation of plastic material and promote its absorption.

In choroiditis, and in exudations in the optic nerve

and retina, mercury is very useful. Inunction is the form I prefer in these cases, associated, if rapidity of effect be desired, with the administration of the iodide of potassium by the mouth.

In order to get satisfactory results by inunction, the blue ointment, or, still better, the ointment of the oleate of mercury, which is equally efficacious and much more pleasant, should be rubbed into the skin in the following manner. About a drachm of the oleate or of blue ointment should be mixed with an equal quantity of vaseline and rubbed into the skin of the inner sides of the thighs and arms and over the chest and groins, every night and morning, choosing a different site each time, and avoiding as far as possible the hairy parts. This ointment should be rubbed in for ten minutes or a quarter of an hour at a time. The skin should not be washed, and the same woollen garment should be worn during as long a period of the time occupied in the inunction as is conveniently possible. The mouth should be carefully cleansed three or four times daily with a chlorate of potash gargle, and the teeth kept clean, and as soon as the spongy line at the margin of the gums is apparent the inunction should be reduced in frequency or stopped.

Whether the effect of the mercury be obtained by means of absorption through the skin or by inhalation of the vapour, as some recent writers assert, I do not pretend to decide. I can only say that by this means, objectionable as it is from a toilet point of view, I have oftentimes seen very good results produced.

In inflammations of the deeper structures, the choroid and retina, the iodides of mercury are useful. They may be given either alone in form of pill or in solution in a mixture containing certain proportions of the iodide of potassium and perchloride of mercury—a useful formula is five grains of the iodide of potassium and half a drachm of the Liq: Hydrarg: Perchlor: to the ounce of water.

In the form of keratitis, which is so often found associated with pegged and notched teeth, cicatrices of ulceration at the angles of the mouth and nose, or more or less flattening of the bridge of the nose (conditions indicative of inherited syphilis), the perchloride of mercury is most useful. Under repeated small doses, associated with small doses of the perchloride of iron, the eye symptoms recede and the health of the patient rapidly improves.

In summing up the value of mercury in the treatment of inflammatory diseases of the ocular structures, I think it may be accepted as an aphorism that wherever there is exudation in diseases of the tunics of the eye mercury is the only really trustworthy remedy.

Another alternative remedy is found very useful in inflammations of the eye — I mean the IODIDE OF POTASSIUM.

It exercises powerful influence on nutrition, and seems to possess the power of promoting the absorption of both serous and plastic exudations. It is, I should say, much less powerful than mercury in dealing with purely plastic exudations, but where the exudation is of mixed character the iodide of potassium is very useful. In all such cases it appears, when used with mercury, to produce more rapid effects than either remedy when used alone.

In the so-called serous iritis, by which is understood that form of inflammation of the uveal tract which is characterized by the presence of products serous and cellular rather than plastic, iodide of potassium is very useful.

In choroiditis, in all the plastic forms, benefit is obtained from large doses of the iodide of potassium, say twenty to thirty grains.

In paralysis of the ocular muscles, and in ptosis due to interference with the action of the nerve supply by pressure from tumours or gummata, or as a result of exposure to cold, iodide of potassium with mercury seems a useful combination. In these cases it is extremely difficult often-times to distinguish the cause of the affection, and to determine the seat of the lesion. Still it is at least safe to presume on the existence of some point of pressure or of neuritis, and to use at least in the early stages large doses of the iodide of potassium as the best known means at command.

THE SALICYLATE OF SODA is a remedy which has come lately into fashion in treating iritis of the so-called rheumatic or relapsing type. Its rapid administration in doses of moderate quantity certainly sometimes seems to relieve pain and cut short an attack, but it is an uncertain remedy and one that at most must be said to be yet on its trial.

IRON is useful as a general remedy in atonic conditions of system, such as frequently accompany phlyctenular keratitis and ulcer of the cornea ; and in tubercular diseases of the iris and choroid I have found the iodide of iron given in full doses for a long period of great value. In diphtheritic conjunctivitis iron administered internally is useful, and in conjunctivitis of erysipelatous character, iron is almost a specific.

In the curious condition known as hemeralopia or night-blindness, which is characterised by inability to see except in strong light, the perception of light being very considerably diminished and the power of adaptation of the eye to reduced light greatly limited (a condition which has also been called torpor of the retina), iron is of great value. In children who present the characteristic symptom of this disease, namely inability to see distinctly directly daylight fades, and who also show the curious dry pearly patches on the conjunctiva, near the corneal margin, the administration of the perchloride of iron causes immediate improvement.

ARSENIC I have found very useful in the cases of children suffering from corneal ulcers and phlyctenular keratitis, especially when given with the wine of iron, and when combined with quinine it is very useful in neuralgia of periodic character.

68

STRYCHNIA is much used in those cases in which it is desired to stimulate the recovery of functional power in the nerves of a part which has been the seat of disease. Thus after an attack of inflammation affecting the optic nerve, or the choroid and retina, when the inflammatory process is over and the exudation has been carried away, strychina is said to be useful. It is generally most effectually used in the form of hypodermic injection, and in increasing doses. By some it is recommended that its use should be begun just before the retrograding of the inflammation is complete.

In the toxic amblyopiæ, and in the loss of vision especially which follows tobacco poisoning, strychnia is said to be useful. It is, however, doubtful how much of the recovery in such cases is to be attributed to its use, and how much to withdrawal of the poison, which is, in all cases, necessary to recovery.

Printed by ROBERT BIRBECK, 313½, Broad Street, Birmingham.